Children Left Home Alone:
Eleven Die in Two Fires
Detroit, Michigan

Investigated by: Mark Chubb

This is Report 070 of the Major Fires Investigation Project conducted by TriData Corporation under contract EMW-90-C-3338 to the United States Fire Administration, Federal Emergency Management Agency.

Department of Homeland Security
United States Fire Administration
National Fire Data Center

U.S. Fire Administration Fire Investigations Program

The U.S. Fire Administration develops reports on selected major fires throughout the country. The fires usually involve multiple deaths or a large loss of property. But the primary criterion for deciding to do a report is whether it will result in significant "lessons learned." In some cases these lessons bring to light new knowledge about fire--the effect of building construction or contents, human behavior in fire, etc. In other cases, the lessons are not new but are serious enough to highlight once again, with yet another fire tragedy report. In some cases, special reports are developed to discuss events, drills, or new technologies which are of interest to the fire service.

The reports are sent to fire magazines and are distributed at National and Regional fire meetings. The International Association of Fire Chiefs assists the USFA in disseminating the findings throughout the fire service. On a continuing basis the reports are available on request from the USFA; announcements of their availability are published widely in fire journals and newsletters.

This body of work provides detailed information on the nature of the fire problem for policymakers who must decide on allocations of resources between fire and other pressing problems, and within the fire service to improve codes and code enforcement, training, public fire education, building technology, and other related areas.

The Fire Administration, which has no regulatory authority, sends an experienced fire investigator into a community after a major incident only after having conferred with the local fire authorities to insure that the assistance and presence of the USFA would be supportive and would in no way interfere with any review of the incident they are themselves conducting. The intent is not to arrive during the event or even immediately after, but rather after the dust settles, so that a complete and objective review of all the important aspects of the incident can be made. Local authorities review the USFA's report while it is in draft. The USFA investigator or team is available to local authorities should they wish to request technical assistance for their own investigation.

This report and its recommendations were developed by USFA staff and by TriData Corporation, Arlington, Virginia, its staff and consultants, who are under contract to assist the USFA in carrying out the Fire Reports Program.

The USFA greatly appreciates the cooperation received from the Detroit Fire Department, particularly Deputy Commissioner Marvin Beatty and other department members who provided valuable information for this report.

For additional copies of this report write to the U.S. Fire Administration, 16825 South Seton Avenue, Emmitsburg, Maryland 21727. The report is available on the USFA Web site at http://www.usfa.dhs.gov/

U.S. Fire Administration

Mission Statement

As an entity of the Department of Homeland Security, the mission of the USFA is to reduce life and economic losses due to fire and related emergencies, through leadership, advocacy, coordination, and support. We serve the Nation independently, in coordination with other Federal agencies, and in partnership with fire protection and emergency service communities. With a commitment to excellence, we provide public education, training, technology, and data initiatives.

TABLE OF CONTENTS

Children Left Home Alone:
Eleven Die in Two Fires
Detroit, Michigan
January-February 1993

Local Contacts: Deputy Commissioner Marvin Beatty
                   Fire Marshal L. Richard Milliner
                   Chief of Arson Charles Evancho
                   Battalion Chief Rodney Parnell
                   Captain Jon Bozich
                   Lieutenant Cohn Prentice
                   Detroit Fire Department
                   250 West Larned Street
                   Detroit, Michigan 48226
                   (313) 596-2900

OVERVIEW

Two Detroit, Michigan, fires are among a group of incidents in the early months of 1993 which raised the Nation's awareness of the problem of unsupervised children. These incidents, which occurred less than a month apart, resulted in the deaths of 11 children under the age of 10. The causes of these fires and the factors which led to the tragic loss of these young lives provide vivid lessons about the dangers of leaving children home alone.

In the first fire, on January 14, a mother left four of her young children alone while she answered a telephone call at a neighbor's home. In the short time she was away, perhaps 10 to 15 minutes, a pot of food on the stove ignited. The fire spread to combustible cabinets and interior finish, growing until it reached flashover. The four children died.

The other fire, February 17, resulted when embers from burning newspapers which were being used to thaw a frozen water line under the wood-frame house, ignited the structure. The fire cut off the only useable exit. Seven children, ranging in age from 7 months to 9 years, lost their lives when they were unable to escape. The front door of the dwelling was secured with bars and a high-security padlock, while all of the windows except one was covered with bars, mesh security grilles, or were otherwise blocked, preventing escape. The window in the room where the children's bodies were found was not blocked by security grilles but was covered by a wood panel door held in-place by a large dresser. The parents of the seven children who died allegedly left the children alone for about 45 minutes while they scavenged abandoned buildings for copper pipe to sell to repay a debt.

1

SUMMARY OF KEY ISSUES

Issue	Comment
Fire Risk	Children under the age of 5 have the highest risk of dying in fire of all groups except adults 75 years of age or older.
Unsupervised Children	Estimates vary widely, but experts generally agree that millions of children are left home alone some of the time.
Fire Causes	Unsupervised children are vulnerable to fires of all causes but are particularly susceptible to fires involving child fire play, juvenile firesetting, and unattended cooking.
Security	Parents who find it necessary to leave children home alone often fear threats to their children's security more than the danger of fire. Even bars, grilles, and special locking arrangements that can be opened from the inside may be difficult for children to operate.
Fire Safety Education	The unique responsibilities of parents and other child caregivers must be addressed in public fire education programs. Juvenile firesetter intervention programs should also be available within the community.
Criminal Prosecution	State criminal statutes set varying standards for neglect and abuse. Parents or guardians whose children die in fires while unsupervised may be subject to prosecution under manslaughter statutes.
Product Design	Efforts are underway to establish Federal standards for child-resistant disposable lighters to reduce the hazards of child fire play at least from that source.

These fires clearly illustrate the tragic consequences which may result when children are left unsupervised. While other factors contributed to both the occurrence of these fires and the inability of the children to escape, the victims of these fires are not unlike millions of children left home alone across the United States. Other cases documented in the press around the time of the two fires in Detroit illustrate that this problem crosses economic and social barriers and affects children in many communities.

December 21, 1992 (Kane County, Illinois) – Emergency dispatchers in suburban Kane County near Chicago, Illinois, were called about a smoke detector activation. A 6-year-old girl left home alone with her 4-year-old sister had gone to a neighbor's house to summon help after water from an overfilled bathtub caused a smoke detector to short circuit. Sheriff's deputies took the girls into protective custody when they learned that the children's parents were vacationing in Acapulco, Mexico, for nine days over the Christmas holidays. When the parents returned to the United States on December 30, they were arrested and charged on 64 counts, including child neglect, child abuse, and child abandonment.

This case received the most attention in the national news media, but a growing number of similar cases fueled a debate, which came to involve the fire service, about what constitutes child neglect, at what age a child is responsible enough to be left without supervision, how long can a child be left alone before it constitutes abandonment, and what should be done about parents who neglect their children in this manner. The Illinois case did not involve a fire, and although the children were in no immediate danger, the fire service was notified in the process of resolving the incident. Other cases have resulted in the deaths of children. In some, injuries and deaths were only narrowly averted.

February 2, 1993 (San Antonio, Texas) – A 13-month-old boy died six days later from injuries he sustained in a fire accidentally set by his 4-year-old brother while playing with a lighter. The 4-year-old was able to escape and notify neighbors who called the fire department. The two boys had been left home alone because their mother, a hotel housekeeper, could not afford day care services.

February 5, 1993 (Milwaukee, Wisconsin) – Firefighters found six children abandoned by their mother when they responded to a fire which started when one of the children put paper in a toaster oven. The oldest child, a 10-year-old girl, told authorities that she had last seen her mother the day before the fire, and that she and her five younger siblings were often left alone for days at a time. Police issued a warrant for the woman's arrest when she could not be located.

February 6, 1993 (Farmington Hills, Michigan) – Five-year-old twin brothers were found locked behind chained doors in separate bedrooms. Police and firefighters were summoned after a delivery driver noticed the two children banging on windows calling for help. Firefighters forced entry through the windows to rescue the children when no one answered the door. The children's mother was running errands at the time and advised police that she had locked the children in their rooms to make sure they stayed there and to keep her ex-husband away from them. Both children were placed in the care of the State Department of Social Services.

February 8, 1993 (Des Moines, Iowa) – Six children were found unsupervised in an apartment with all of the burners on the stove turned on. The children were found, unsupervised for hours, after a babysitter had left them alone. The bathroom was also locked.

February 9, 1993 (Carteret, New Jersey) – A boy and girl, ages 4 years and 17 months, were found unconscious by firefighters responding to their home for a fire caused by unattended cooking. A neighbor summoned the fire department when she heard one of the children calling "Daddy! Daddy!" and saw smoke pouring from the apartment. The children's parents had left the kids home alone while making a trip to a nearby supermarket.

March 1, 1993 (Miami, Florida) – A 6-year-old disabled girl died in a fire started by her 5-year-old brother. The two children and their 3-year-old sister were left alone for about 20 minutes while their mother was at a nearby coin laundry. The boy set a living room sofa on fire while playing with matches. A smoke detector above the sofa activated, alerting a neighbor who rescued the boy and notified the fire department. Firefighters rescued the 3-year-old and found the 6-year-old on the living room floor but were unable to revive her.

March 8, 1993 (Houston, Texas) – A 4-year-old boy left home alone with his 4-month-old sister died after setting fire to clothing in a closet. The 4-month-old girl was rescued by firefighters and was hospitalized and treated for smoke inhalation. Two older children, ages 8 and 11, were at school when the fire occurred. The children's mother reportedly left the youngsters alone while running an errand. Investigators believe she was gone for approximately one hour when the fire was discovered by neighbors. Investigators revealed that the boy had set other fires in the past but had not received counseling for his firesetting behavior.

According the Harris County Children's Protective Services authorities, the woman was on probation for a drug-related conviction at the time of the fire. While she was incarcerated as a result of her arrest, the three oldest children were in protective custody.

March 13, 1993 (Toledo, Ohio) – Two children, ages 9 and 10, died in an early morning house fire which started when an overturned electric space heater ignited their bedding. When the children were awakened by the fire, they tried to smother it with clothing and then sought refuge in a closet. Their 11-year-old brother, also in the room when the fire started, was burned but managed to escape and call the fire department. The children's mother had left them alone overnight, and she now faces involuntary manslaughter and child endangerment charges.

THE DETROIT FIRES

The two fires in Detroit, Michigan, January 14 and February 17, 1993, illustrate the danger of leaving children alone even for relatively short periods of time. The first case is an example of what can happen in a moment of distraction or inattention. The second fire illustrates the consequences of lack of parental supervision when other fire safety problems exist.

The Sheridan Street Fire

On the afternoon of January 14, a young mother was called to a neighbor's home to answer a telephone call concerning her 5-month-old son, who was in critical condition in the children's intensive care unit of a local hospital after receiving an organ transplant. She ran to the neighbor's house to take the call, leaving her other four children – ages 2 through 5 – alone in the house. She also left a pan of food cooking on the stove.

Less than 15 minutes later the kitchen of this single-family home was in flames. Would-be rescuers from nearby houses were unable to enter the house to reach the children. When firefighters arrived, the fire had already spread to other rooms on the first floor of the home and was venting from the first floor windows and doors.

Although the fire was controlled relatively quickly, the damage was already done. The four children were found dead on the second floor.

The Mack Street Fire

The afternoon of February 17, was a particularly cold one in Detroit. Temperatures had plummeted overnight and hovered in the low teens during the day with high winds and occasional snow showers.

The house at 2258 Mack Street was a 1 1/2-story wood-frame structure built in the early 1870s. It was located in an economically-depressed neighborhood on the city's east side; the lots on either side of the house were vacant.

The cold temperatures and lack of adequate insulation in the building's crawl space beneath the house caused pipes under the kitchen sink to freeze overnight. Sometime in the early afternoon, the adult male resident of the home opened an access panel on the side of the house, entered the crawl space, and began attempting to thaw the frozen pipes. To do this, he fashioned torches from rolled newspapers and lit them on fire. The lit newspapers were then held close enough for the flames to contact the water pipes, which were fastened to the floor joists.

While he was working under the house another man drove up to the house, walked up to the access panel, and asked the first man to come out to discuss an unpaid debt. The second man offered to help the first man and his wife collect money to replay the debt by taking them to salvage copper pipe from some nearby abandoned buildings. A witness, who was in the area at the time, told investigators that she saw the man and his wife leave with the second man in his automobile. The children were left alone inside the house.

According to witnesses, the seven children, ranging in age from 7 months to 9 years, were often left alone for extended periods of time by their parents. Although three of the children were of school age, on the day of the fire, a Thursday, all of the children were at home when at approximately 2:00 p.m. passersby noticed smoke coming from the building and called the fire department.

Firefighters from Engine 6 and Ladder 10, located about 10 blocks away, responded with two other engine companies, another truck company, a rescue squad, and a battalion chief. When Engine 6 and Ladder 10 arrived on the scene two minutes after the alarm was transmitted, fire was venting from the windows on the southwest side and rear of the house and thick smoke was pouring from the other openings. See Appendix A for the floor plan of the house.) A bystander told arriving firefighters that children were trapped inside – possibly in the living room, confirming information transmitted to them over the alarm printer in the station when they were dispatched.

They tried to enter the house through the front door but found it secured with a heavy iron gate and high-security padlock. Of the two front windows of the house, the east window was covered with a security grate and the west window was obstructed by bicycles and other items stacked in front of it on the inside. Similarly, the windows on the southwest side of the house were covered by bars or security grates. On the northeast side of the house, only the window at the north end was not covered by bars or grates, but it was blocked by a panel door placed over it on the inside. As a result of these security devices and obstructions, firefighters were forced to enter the house through the rear doorway right where the fire had started and was most intense.

Firefighters from one company advanced through the kitchen and dining room to the living room to begin searching where the children were last seen by bystanders. Another crew fought through the kitchen and bathroom to begin searching the attic. Within minutes, firefighters had brought the fire under control and discovered the bodies of the seven children in the front bedroom, located just off of the living room.

Investigators believe that the children, led by the 9-year-old, sought refuge in the front bedroom because it was remote from the fire. Most of the windows were blocked or locked, and the back door in the kitchen was blocked by the fire. Although the bedroom sustained almost no fire damage, it could not provide refuge from the choking smoke produced when the fire reached flashover in the kitchen.

The parents returned approximately 20 minutes after firefighters arrived on the scene. Although fire investigators had probable cause to arrest the parents for child neglect and involuntary manslaughter, they were taken by Detroit EMS to the hospital where the children were transported from the fire scene. There the parents identified the bodies and were taken into custody by Detroit police officers who had determined that the woman was wanted on an outstanding arrest warrant. Detroit Police Department homicide detectives held the parents for almost 48 hours for questioning before releasing them for their children's funerals.

FIRE RISK

To understand the fire risks associated with leaving children unsupervised, it is important to understand how common it is for children to be left home alone. The large number of children in self-care arrangements must then be weighed against the relative risk of fire deaths and injuries among children and the causes of the fires which result in those injuries and deaths.

Children in Self-Care

A number of studies in the past decade have focused on the special problems of latchkey children. The practice of leaving children home alone during the period between the end of the school day and the return of one or both parents from work – often two or three hours – is common in the United States. The practice of leaving even younger children home alone or in the care of children under 15 years of age is also prevalent. Studies by the U.S. Bureau of the Census (O'Connell and Bachu, 1992) and the former Department of Health, Education and Welfare (now the Department of Health and Human Services and Department of Education) present widely different estimates of the actual number of children in self-care arrangements, but experts have concluded that the number is probably in the millions, possibly approaching 10 million youngsters under the age of 13. One study by the Child Welfare League of America (Kraizer, Whitte, and Fryer, 199) places the number of children between 5 and 9 years of age left home alone as high as 8 million, and when older children are included, the number rises to 27 million.

National estimates of the number of young children, under 5 years of age, who are left alone or in the care of older siblings are not available. However, a sense of the problem can be obtained by considering one community, Harris County, Texas. Here, in 1993, Children Protective Services (in Houston) investigated 1,942 cases of unsupervised preschoolers (each family is a separate case regardless of the number of complaints or number of children in the household).

The increase in the number of single-parent households and families where both parents work full-time is almost certainly driving up the numbers of home alone youngsters.

The impact of this trend is unclear. Bethesda, Maryland, Clinical Child Psychologist Thomas Long, interviewed for an article in *Time* Magazine (Willwerth, 1992), explains that latchkey children – those who routinely care for themselves part of the day – fall into two groups: "those who see themselves as independent and capable, and those who see their situation as one of rejection and abandonment."

Considerable consensus exists in the child psychology and social welfare literature regarding the absolute inadvisability of leaving preschool-age children home alone. Most experts cite accident and injury statistics and situations like the ones detailed in this report to substantiate their conclusions (Peterson, 1989). Guidance on the subject of leaving older children home alone in latchkey situations – to care for themselves after school until a parent returns home – varies but generally converges on the point that considerable preparation is required even in cases where children seem responsible and mature (Coleman, Johnson, and Todd, 1990; Peterson, 1989). Peterson bolsters this argument with research suggesting that parents and children alike consistently overestimate the children's abilities to deal with hazardous situations.

Relative Fire Risk

The impact of the child care dilemma on fire risk becomes clearer when the fire death and injury statistics compiled through the National Fire Incident Reporting System (NFIRS) are analyzed. The relative risk of dying in a fire is greatest among the very young and the very old. Only adults over the age of 75 have a higher relative risk of dying in a fire than children under 5 years of age (U.S. Fire Administration, 1990). (See Table 1 and Figure 1.)

Figure 1. Relative Risk of Fire Death by Age – 1990

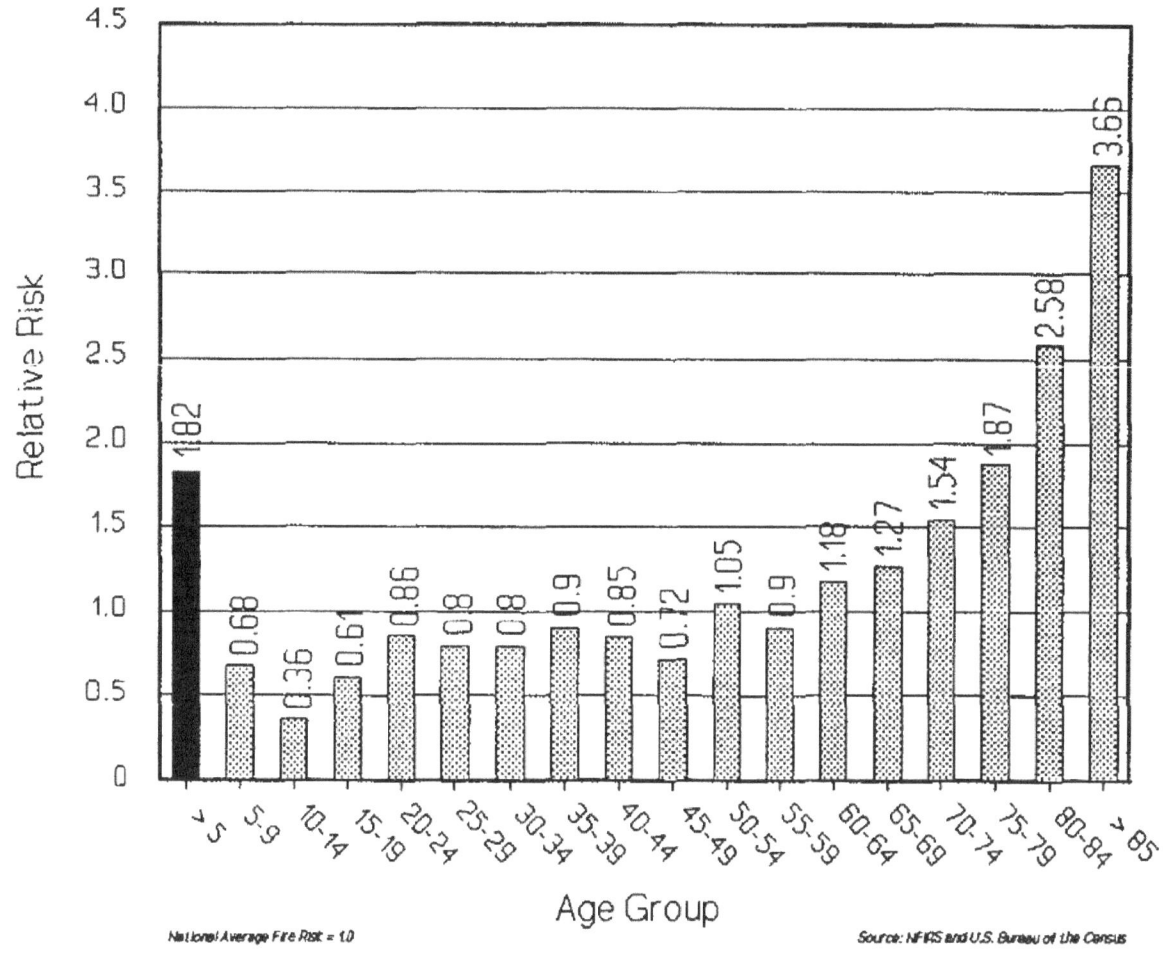

National Average Fire Risk = 1.0

Source: NFIRS and U.S. Bureau of the Census

Fire Death Causes

The leading cause of fire deaths for children under the age of 5 is children playing with matches, lighters, and other open flame devices. Incendiary and suspicious causes rank third just behind heating (U.S. Fire Administration, 1990). Like the fire risk statistics, it is unknown how many of these cases involve unsupervised children. However, failure to keep matches and lighters out of the reach of children usually involves some degree of inadequate supervision. Some of the fires listed as incendiary and suspicious may be caused by child firesetting, too. Set fires and fires caused by children playing can be impossible to distinguish, particularly when no one survives.

In the two cases discussed in this report where cooking was responsible for fires, the parents had been cooking but left a pot unattended or were called away for a short period of time. These fires simply underscore the dangers of unattended cooking in all environments but serve to illustrate the dangers to children when they are inadequately supervised even for short periods of time.

Table 1. Relative Risk of Fire Death by Age Group

Age Group (Years)	Deaths	Population (1,000s)	Relative Risk
Less than 5	713	18,874	1.82
5-9	255	18,064	0.68
10-14	130	17,191	0.36
15-19	227	17,790	0.61
20-24	346	19,305	0.86
25-29	357	21,356	0.80
30-34	367	21,990	0.80
35-39	374	20,031	0.90
40-44	313	17,814	0.85
45-49	206	13,826	0.72
50-54	247	11,368	1.05
55-59	196	10,473	0.90
60-64	260	10,618	1.18
65-69	265	10,077	1.27
70-74	257	8,021	1.54
75-79	239	6,145	1.87
80-84	211	3,932	2.58
85 and older	232	3,050	3.66
TOTALS	**5192**	**249,925**	**1.00**

Source: NFIRS and U.S. Bureau of the Census

Juvenile Firesetting Behavior

With juvenile firesetting such a prevalent cause of fire deaths and injuries among young children, this fire cause deserves special attention.

The February 2 fire in San Antonio, Texas, and the March 1 fire in Miami, Florida, are examples of the problems and consequences of juvenile firesetting by unsupervised children. The San Antonio fire was started by a 4-year-old boy playing with a lighter and killed his 13-month-old brother. The boys' mother was at work at the time of the fire. In the Miami fire, a 6-year-old disabled girl died when her 5-year-old brother lit the living room sofa on fire while playing with matches. Their mother was out at a nearby coin laundry.

Table 2 describes six characteristics of juvenile firesetters. It seems likely that the feelings of low self-esteem, anger, or aggression described may be fostered by parental abuse, neglect, or absence.

REMEDIES

Many communities have proactive programs to educate children, as well as the public at-large, about fire safety in the home. Some communities have implemented juvenile firesetter intervention programs. Local fire officials have also spurred changes at the National level by pressuring manufacturers and the Federal government to address the dangers posed by disposable lighters. In cases where child abuse or neglect is suspected, some communities have sought criminal charges against parents and guardians. All of these approaches contribute to managing the fire risks of unsupervised children but don't solve the problem.

Table 2. Juvenile Firesetter Profile

- **Demographic:** Usually male; between ages 3 and 13, firesetting is often motivated by curiosity, in 14-18 year olds, behavior is more complex and severe.

- **Physical:** Many firesetters exhibit unusual incidence of physical illnesses, especially allergies and respiratory problems, and higher than average incidence of accidents involving injury. Children who have been physically or sexually abused may be at higher risk of being juvenile firesetters.

- **Cognitive:** Most juvenile firesetters test normal in verbal and intellectual skills, but often exhibit higher than normal incidence of learning disabilities. Poor academic performance resulting from these learning disabilities may be accompanied by conduct and behavior problems that become worse as the child grows older.

- **Emotional:** Firesetting behavior may be used as an expression of anger or aggression.

- **Motivational:** Firesetting may arise from curiosity and a need to interact with surroundings, may be an outlet for anger and aggression stemming from emotional conflicts, or may come from a desire for recognition or acceptance.

- **Psychiatric:** Most frequent diagnosis of juvenile firesetters is conduct disorder. Low self-esteem, low frustration tolerance resulting in irritability or temper outbursts, poor academic achievement, and behavior difficulties at home or at school. Higher than average diagnosis of hyperactivity is also common.

Source: J. Gaynor, "Juvenile Fire Setting," in **Human Fire Related Behavior Course Guide**, National Fire Academy Open Learning Fire Service Program, 1989, pp 4-9 – 4-13.

Criminal Statutes

Several of the cases cited in this report came to public attention because of criminal proceedings against parents accused of child neglect or abuse. However, little consensus has emerged about when a parent's absence is sufficiently negligent to constitute criminal conduct.

On March 1, the Wayne County Prosecutor John O'Hair filed manslaughter charges against the parents of the seven children who died in the Mack Street fire in Detroit, characterizing the actions of the parents as "gross negligence." The attorney for the couple responded that it was unfair to prosecute the parents, "when they haven't had a chance to deal with their grief." A jury must ultimately decide whether a crime occurred, and if so, whether the loss of their children is penalty enough.

One of the important things the jury must decide is whether the parents' conduct constituted child neglect. Michigan law does not establish a minimum age when children are considered sufficiently responsible to legally care for themselves. In the past, cases have been decided on individual merit. Other States have filled this gap by specifying minimum ages (Peterson, 1989). The ambivalence about minimum age standards for child self-care is in sharp contrast to the consensus which exists in other areas of the law regarding the relationship between age and responsibility, such as the minimum drinking, driving, and voting ages.

The legal ground surrounding the presumption of criminality in cases of unsupervised children is so shaky that there is little agreement about the merits of prosecution even when a child's death results. The decision whether or not to prosecute in a given situation is further complicated by political considerations, especially in communities where the prosecutor is elected to office. Nonetheless, when charges are filed against parents or guardians, they usually allege involuntary manslaughter, criminal negligence, child neglect, child abuse, or some combination of these charges. In most States, these charges constitute serious misdemeanors or felonies punishable by substantial fines, prison sentences, or both.

Public Fire Safety Education

Public fire safety educators have focused on the hazards of children playing with fire and juvenile firesetting behavior for some time. Many departments have developed and implemented programs to deal with these serious problems. The USFA has published the following resource materials to assist local agencies in implementing such programs:

- Curious Kids Set Fires (5-0121), 1988 (a kit of ready-to-use public education materials)

- Preadolescent Firesetter Handbook, Ages 0-7 (FA-83), 1988

- Preadolescent Firesetter Handbook, Ages 7-13 (FA-82), 1988

- Adolescent Firesetter Handbook, Ages 14-18 (FA-80), 1988

- Public Education Today: Fire Service Programs Across American (FA-98), 1990

Other approaches to public fire safety education address fire risk among children, including unsupervised children. Like many other cities, Detroit is in the midst of a massive campaign to promote the installation of home smoke detectors. The department is targeting its efforts in the neighborhoods with the highest fire deaths rates. Fire companies are distributing free smoke detectors and installing the devices in homes where they are needed and occupants are otherwise unable to put them up themselves. The department expects to distribute as many as 15,000 free smoke detectors through this program. Ironically, this campaign was kicked off just four days before the fatal fire on Mack Street.

Unfortunately, smoke detector campaigns, like many other approaches, fail to address the key problem of unsupervised children. All of the incidents discussed in this report occurred during daytime hours, when most of the occupants would be awake. Although smoke detectors played a role in at least one of the incidents – the Miami fire in which a neighbor alerted to the fire by a detector rescued the 5-year-old who started the blaze – detectors are designed primarily to alert people who are asleep when a fire starts. The outcome of fires involving hazards which produce fast flaming fires, such as cooking, child fire play, or child firesetting behavior, may not be significantly altered by smoke detectors, especially if children are not adequately trained to respond.

Programs for educating children about the dangers of fire and equipping them with the skills to successfully escape a fire should begin as early as possible. Peterson (1989) suggests that active role-playing, repetition, and rewards are effective techniques for reinforcing fire safety behaviors, even among preschool age children. If a 3-year-old can start a fire, then that same child is old enough to understand how to prevent one as well. The types of skills needed to prepare youngsters for the dangers of fires should be carefully identified. At a minimum, every child should understand how to escape a fire, how to call for help in the event of emergencies, and how to reach a parent or other responsible adult.

Currently, much of the emphasis in fire safety education is on reaching elementary-school-age children. However, this approach may leave preschoolers and others at high-risk, such as children with poor academic skills or learning disabilities and those who are frequently absent from school. A once-a-year school fire safety program runs the risk of missing these children. Fire safety programs for adults and for children should make clear the dangers of leaving children home alone and of leaving children in charge of even younger children. It may also be necessary for the fire department to build coalitions with other organizations and agencies in their communities who address the social welfare of families and children. Such cooperative efforts should be aimed at educating parents about the dangers of leaving children home alone and expanding before-school and after-school programs for latchkey children.

Product Design Standards

Recently, at the request of several local and State fire service organizations, Polyflame Concepts USA and Philip Morris USA issued a voluntary recall of a novelty lighter deemed inappropriate and unsafe due to its lack of child-resistant features and its attractiveness to young children. Meanwhile, the U.S. Consumer Product Safety Commission (CPSC) has announced its intention to propose new Federal standards to require that novelty and disposable lighters be designed with child-resistant features. These standards are expected to affect the design of 95 percent of the nearly 600 million lighters sold in the United States annually. Officials hope it will lead to a dramatic reduction in fires, fire deaths, and fire injuries caused by juvenile fire play (*The Gated Wye*, 1993). Other CPSC actions, relating to kerosene heaters, the flammability of children's sleepwear and of mattresses, and the promotion of smoke detectors have contributed to home fire safety as well.

ANALYSIS

It is unsafe to leave young children alone for even short periods of time. Children in the fire incidents described in this report were left alone for periods ranging from 15 minutes to several days. Although the cases of long-term neglect have caused the most controversy, the fires described above show that even commonplace situations involving short periods of parental absence can be catastrophic. A parent need only be absent long enough for a child to strike a match or flick a lighter. In addition, the family environment a child experiences may have an impact on fire risk, particularly in the case of juvenile firesetting behavior. Feelings of neglect and emotional abuse wrought by parental absence may produce behavioral problems which result in firesetting.

Traditional fire safety education may fall short of addressing these problems. Preschoolers, children with learning disabilities, and those who miss school often may be difficult in reach with traditional school-based fire safety education programs. In addition, many children are forced to take responsibility for younger siblings. Few fire safety programs today warn of the danger this practice creates.

A coordinated approach to managing the fire risk of unsupervised children appears warranted. Fire risk is only one aspect of the problem of unsupervised children in our society and part of the National child care dilemma. The scarcity of affordable day care and growing number of single-parent households and two-income families suggest that these problems will become worse before they get better.

A successful strategy for dealing with fire risk must recognize the roles played by the non-fire agencies and organizations in the community. Schools and social welfare agencies play important roles in reaching children in at-risk families. Fire departments would do well to ally themselves with these caring professionals. Such alliances are likely to suggest more ways the fire service can address this very serious problem.

LESSONS LEARNED

1. **Leaving small children alone for any period of time is unsafe.**

   The danger of a child dying in fire is not related to the duration of parental absence or intent. Well-meaning, conscientious parents can lose their children in a moment of inattention or neglect. Steps to minimize the risk of childhood injury or death in fire must begin with parents and guardians by making them aware of the risks associated with leaving children home alone.

2. **The fire department's fire safety education programs, for both children and adults, should include messages about the fire risks of children left home alone and of children left alone to care for even younger children.**

   Parents, the community at large, and children themselves all have a stake in this important area of fire safety and have a right to this information. Just as school children have carried the message of the importance of smoke detectors home to their parents, they hopefully will do the same with the "home alone" fire safety messages addressed in this report when they are exposed to them at school or at other types of fire department presentations.

3. **Efforts to educate children about the risks of fire can start as early as preschool.**

   Children as young as 3 years of age have started fires while playing with matches and lighters. If a child is old enough to start a fire, he or she may be old enough to learn how to prevent and escape one as well. Young children are attracted to fire by their innate curiosity. Early childhood fire safety education should try to satisfy the curiosity of young children while pointing out the dangers of fire.

4. **Any community that has not already one so should consider establishing a juvenile firesetter intervention program.**

   Juvenile firesetter intervention programs can be an integral part of a proactive approach to preventing tragedies like those described in this report. Many children become involved in fire play and firesetting long before their behavior results in such tragedies. Early identification and appropriate intervention and counseling can prevent a tragedy by helping parents and children work through the difficulties that are causing the child to experiment with fire.

5. Uniform standards of criminal conduct should be developed to enable prosecution of negligent parents.

   Fires resulting in serious injury or death to children should be treated like other negligent acts which lead to accidental death. Motor vehicle statutes developed in response to Federal funding guidelines have made causing a fatal drunken driving accident a distinct criminal offense in most States. However, uniform standards of care for the supervision of children do not exist. As a result, lack of such standards creates a murky environment for prosecuting neglectful parents when their children are killed or seriously injured in fires.

6. Products should be designed to minimize risks to children.

Federal efforts to establish safety standards for the design of disposable lighters should be a priority. The fire service and fire safety educators should become actively involved in the discussion to establish these standards. Their input and guidance has already been an important factor in removing other hazardous consumer products from the marketplace.

7. The fire service should seek out alliances with schools, social welfare agencies, and other interested parties to minimize the fire risk of unsupervised children.

Many of the agencies involved in child welfare would welcome support and input from the fire service. Likewise, their support of fire service programs will be beneficial to fire safety education programs aimed at at-risk children. A surprising array of new approaches to the fire risk problem may be opened up by entering into such alliances.

REFERENCES

Bandow, D., "Home Alone: Should Congress Play with Family Leave," *Business and Society Review*, **77**, Spring 1991, p. 41.

Bernstein, L., "Home Alone," *Woman's Day*, **55** (14), Sept. 1, 1992, p. 117.

"Child-resistant lighters." *The Gated Wye*, **110**, Salem, Ore.: Oregon State Fire Marshal, Feb. 1993, p. 1.

Coleman, M., C.E. Johnson, and C. M. Todd, "Self-Care and School-Age Child Care Resources," *Journal of Home Economics*, **82** (1), 1990, pp. 53-56.

Comer, J. P., "11 through 13: Staying Home Alone," *Parents*, **67** (10), Oct. 1, 1992, p. 221.

Edger-ton, R. B. and C. H. Kempe, *Child Abuse and Neglect: Cross- Cultural Perspectives*, Berkeley, Calif.: University of California Press, 1981.

Foley, D., "A Guide for Latchkey Kids," *Woman's Day*, **53** (13), Sept. 4, 1990, p. 120.

Gaither, N. E., The effects of a 4-H survival skills program on decisionmaking skills of latchkey children, M.A. Thesis, University of Maryland, 1985.

Gaynor, J., "Juvenile Fire Setting," in *Fire Related Human Behavior Course Guide*, Emmitsburg, MD: National Fire Academy Open Learning Fire Service Program, 1989.

Heins, U. R., Variables contributing to fear in latchkey children, M.A. Thesis, University of Maryland, 1984.

Institute for Judicial Administration-American Bar Association Commission on Juvenile Justice Standards, Standards Relating to Abuse and Neglect Recommendations by the IJA-ABA Joint Commission on Juvenile Justice Standards, Cambridge, Mass.: Ballinger Publishing, 1981.

Koblinsky, S. A., and C. M. Todd, "Teaching Self-Care Skills to Latchkey Children: A Review of Research," *Family Relations*, **38** (4), October 1, 1989, p. 431.

Kraizer, S. S. Witte, and G. E. Fryer, Jr., "Children in Self-Care: A New Perspective," *Child Welfare*, **69** (6), Nov. 1990, p. 571.

Larner, M. and A. Mitchell, "Meeting Child Care Needs of Low-Income Families," *Child and Youth Care Forum*, **21** (5), Oct. 1992, pp. 317-334.

Long, L. and T. Long, *The Handbook for Latchkey Children and Their Parents*, New York: Arbor House, 1983.

Lovko, A. M. and D. G. Ulman, "Research on the Adjustment of Latchkey Children: Role of Background/ Demographic and Latchkey Situation Variable," *Journal of Clinical Psychology*, **18** (1), March 1, 1989, p. 16

National Center on Child Abuse and Neglect, Interdisciplinary Glossary on Child Abuse and Neglect: Legal, Medical and Social Work Terms, Washington, DC: U.S. Department of Health, Education, and Welfare.

O'Connell, M. and A. Bachu, "Who's Minding the Kids? Child Care Arrangements: Fall 1988," *Current Population Reports*, p. 70-30, Washington, DC: U.S. Bureau of the Census, 1992.

Peterson, L., "Latchkey Children's Preparation for Self-Care: Overestimated, Under-rehearsed and Unsafe," *Journal of Clinical Child Psychology*, **18** (1), 1989, pp. 36-43.

Robinson, B. E., and others, *Latchkey Kids: Unlocking Doors for Children and Their Families*, Lexington, Mass.: Lexington Books, 1986.

U.S. Bureau of the Census, Primary Child Care Arrangements Used by Employed Mothers for Children Under 15, by Age of Child: 1986-1987, Table No. 623, *Statistical Abstract of the United States*, 111th ed., Washington, DC: U.S. GPO, 1991, p. 377.

U.S. Fire Administration, *Fire in the United States*, 7th ed., Emmitsburg, MD: Federal Emergency Management Agency, 1990, pp. 11-12, 224-225.

Willwerth, J., "Society: Helping the Home Alone Set," *Time*, **141** (9), March 1, 1993, p. 46.

Wright, R., "Who's Watching the Children?" *McCall's*, **117** (4), Jan 1, 1990, p. 22.

Floor Plans for 2258 Mack Street, Detroit, Michigan

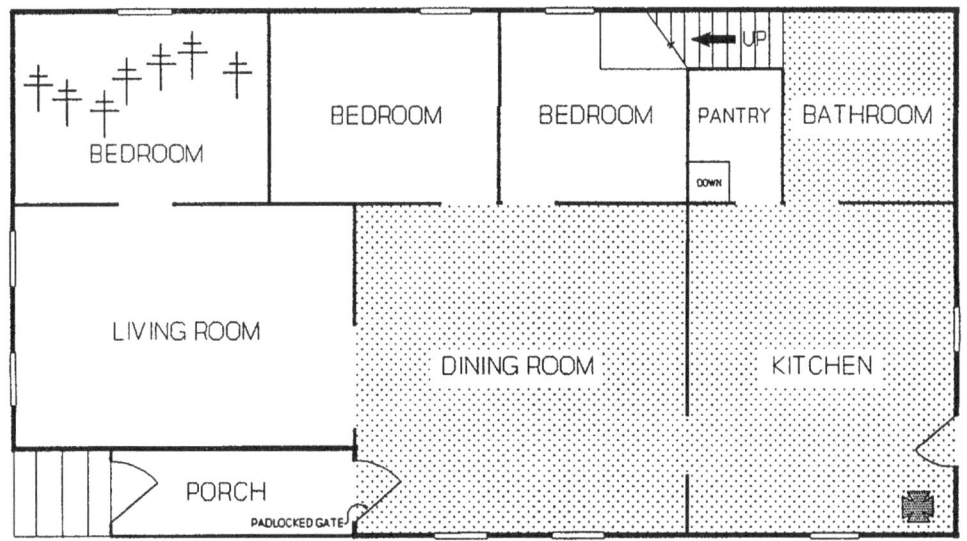

First Floor
2258 Mack Street
Detroit, Michigan

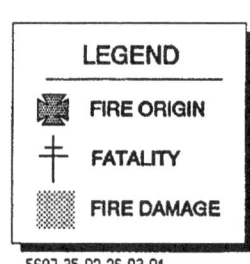

Appendix A (continued)

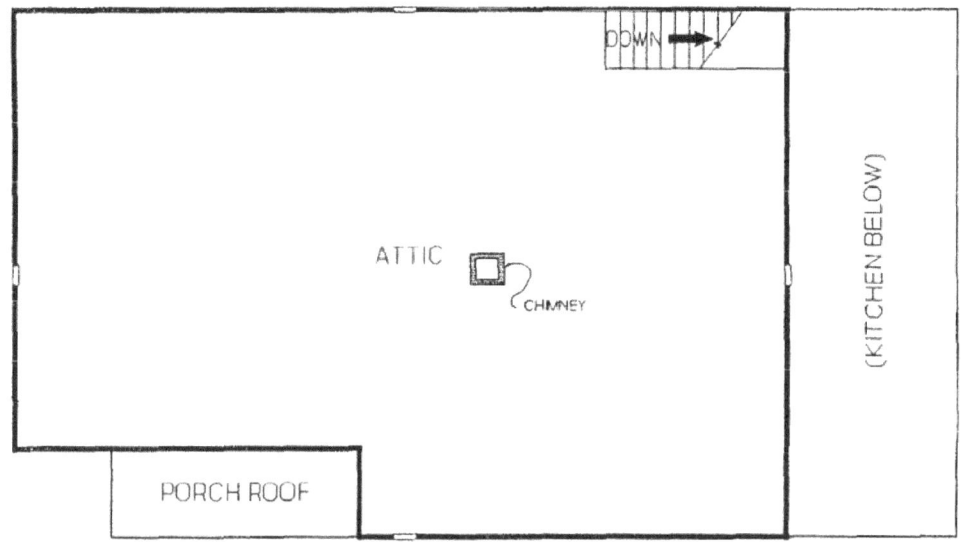

Second Floor
2258 Mack Street
Detroit, Michigan

5697-35-02-26-93-02

APPENDIX B

Photographs of 2258 Mack Street, Detroit, Michigan

Photo by Detective/Sergeant Joel DeKraker

Exterior of 2258 Mack Street from the front. The front door was secured by an iron gate with a heavy padlock. The left window was covered with security bars.

Appendix B (continued)

Photo by Detective/Sergeant Joel DeKraker

Exterior view from the rear of 2258 Mack Street looking toward the point of origin. The fire started in the crawl space under the corner of the house.

Appendix B (continued)

Photo by Detective/Sergeant Joel DeKraker

Exterior view of typical window security grille. This one was installed over the bathroom window. Similar ones were installed on the windows of the center and rear bedrooms.

Appendix B (continued)

Photo by Detective/Sergeant Joel DeKraker

**Burglar bars removed from the rear kitchen window. Similar bars were
installed on the dining room windows and one of the
living room windows.**

Appendix B (continued)

Photo by Detective/Sergeant Joel DeKraker

Interior view of the iron gate at front door. Note the padlock near the
center of the photograph. The door frame and wainscoting are heavily
charred after the dining room reached flashover.

Appendix B (continued)

Photo by Detective/Sergeant Joel DeKraker

Interior view of living room looking toward the entrance to the bedroom where the children's' bodies were found by firefighters. One of the living room windows was covered with a security grille; the other was blocked by bicycles and other debris.